FLORA OF TROPICAL EAST AFRICA

PAPAVERACEAE

G. Ll. Lucas

(East African Herbarium)

Annual, biennial or perennial herbs, rarely shrubs, and only 1 tree genus (*Bocconia* L.), with white, yellow or orange coloured latex. Leaves alternate or rarely whorled, exstipulate, entire to much divided (palmately, pinnately etc.). Flowers usually solitary, conspicuous and large, bisexual, regular, hypogynous. Sepals 2–3, imbricate, usually free or calyptrate, caducous. Petals (4–)6(–12), more rarely absent, imbricate, arranged in 1–2(–3) whorls, crumpled in bud. Stamens free, usually numerous, spirally arranged, rarely 4 and cyclic ; anthers 2-celled with longitudinal dehiscence. Ovary superior, usually unilocular, more rarely with 2 to several locules ; ovules numerous ; placentation parietal. Stigmas opposite or alternate with placentas. Fruit usually a capsule dehiscing by valves or pores, rarely indehiscent. Seeds small, numerous, with minute embryo and copious, usually oily, endosperm.

There are no indigenous species of *Papaveraceae* present in tropical East Africa, but *Argemone mexicana* L. is a well established alien. It must also be remembered that many species of Poppy (*Papaver* L.) are cultivated as ornamentals and there is at least one record where a species has escaped and become very locally established. This is *Papaver rhoeas* L. which was found in Tanganyika at Mkusi near Lushoto, on 31 Aug. 1950 (*Verdcourt* 339).

ARGEMONE

L., Sp. Pl. : 508 (1753) & Gen. Pl., ed. 5 : 225 (1754) ; Prain in J.B. 33 : 130 (1895) ; Ownbey in Mem. Torr. Bot. Club 21(1) : 21 (1958)

Annual or biennial herbs, sometimes perennial and shrubby. Stems glaucous, erect to 1·5 m. high, unarmed to prickly. Leaves glaucous, sessile, pinnately lobed or incised with a dentate prickly margin, and often the surface prickly. Flowers large, terminal or cymose. Sepals (2–)3(–6) with a subterminal terete horn ; outer surface usually sparsely prickly. Petals 6, normally in 2 whorls of 3, usually white or yellow. Anthers basifixed ; filaments filiform. Carpels (3–)4–6(–7) ; stigmatic lobes same number as carpels and opposite the placentas. Capsule opening by means of valves. Seeds brown-black, up to 3 mm. in diameter ; embryo larger than in other genera, i.e. about 2/3 seed-length.

A. mexicana *L.*, Sp. Pl. : 508 (1753) ; Oliv., F.T.A. 1 : 54 (1868), Prain in J.B. 33 : 308 (1895) ; Fedde in E.P. IV. 104 : 273 (1909) ; Ownbey in Mem. Torr. Bot. Club 21(1) : 29, fig. 2 (1958). Type : specimen from Central America (LINN)

Herbs erect up to about 1·2 m. high, often branching near the base. Leaves sessile, ± sheathing (amplexicaul) ; prickles if present mainly on the

1

FIG. 1. *ARGEMONE MEXICANA*—**1,** flowering branch, × 2/3 ; **2,** flower with two petals removed, × 1 ;
3, stamen, × 6 ; **4,** top of dehisced capsule showing vascular strands and persistent stigma, × 2 ;
5, seed, × 8.   1, from *Haarer* 969 ; 2-5, from *Wallace* 1299.

veins of the abaxial surface ; leaves with midrib and veins irregularly
outlined in greyish-white on the adaxial surface. Flowers 2·5–4·5 cm. in
diameter, subtended by 1 or 2 foliaceous bracts. Sepals 3, caducous, covered
to a greater or lesser degree with prickles. Petals 6, yellow more rarely
cream-white, glabrous. Stamens numerous. Stigma and style persistent,
1–3 mm. long in fruit. Capsule oblong to broadly ellipsoid, 1·2–4·5 × 2·5
× 2·0 cm. ; valves 4–6, which split away from the vascular strands for
about 1/3 of their length, exposing a cage-like frame of these vascular
strands, attached apically to the persistent stigma ; outer valve-surface
prickly in all tropical East African specimens seen (see Note below). Seeds
up to 2 mm. in diameter, subspherical with a small beak at one end ; testa
pitted in rows radiating from the micropyle end. Fig. 1.

UGANDA. Ankole District : Kamatalizi, 26 Oct. 1950, *Jarrett* 199 ! ; Mengo District :
   Kalolo nr. Kampala, 20 Sept. 1932, *Lab. staff in Tothill* 1181 !
KENYA. Kiambu District : Kabete, 14 Nov. 1961, *Verdcourt* 3230 ! ; Kilifi District :
   Malindi, Oct. 1951, *Tweedie* 1059 ! ; Tana River District : Garissa, 25 Dec. 1942,
   Bally 1980 !
TANGANYIKA. Mwanza District : Ukerewe I., *Conrads* 959 ; Arusha, Dec. 1927,
   Haarer 969 ! ; Tabora, Mar. 1939, *Lindeman* 592 !
ZANZIBAR. Zanzibar I. : without locality, Sept. 1860, *Speke* ; Tanguu Road, 29 Aug.
   1930, *Vaughan* 1463 !
DISTR. **U**2, 4 ; **K**4, 7 ; **T**1–4, 6, 8 ; **Z** ; a native of Central America introduced as a
   weed in East Africa as in most other tropical and subtropical regions of the world.
HAB. Waste places, roadsides and abandoned cultivated ground ; 0–1800 m.

NOTE. In West Africa many specimens have none or very few prickles on the stem,
   leaves and capsule and this is treated by Ownbey as forma *leiocarpa* (Greene) G. B.
   Ownb.

INDEX TO PAPAVERACEAE